How to Diagnose and Repair Carburetor Problems

I0504859

A Complete Guide to Help You Troublehoot Carburetor Problems

Gladys E. Omo

Contents

INTRODUCTION

Prior to the advancement of electronically controlled engine management systems, which came about in the late 70's, fuel and air was conveyed into the engine of a vehicle mechanically through a component of the vehicle known as the carburetor.

A Carburetor is a mechanical device that mixes fuel and air into suitable combination that can be ingested for internal use by an engine. Today's modern vehicles are no longer built with carburetors, however they are still commonly found on many older cars and trucks, as well as some classic vehicles.

The carburetor uses a component called the "intake vacuum" to deliver fuel to the engine. As air is pulled down through the throat of the carburetor by the intake vacuum, so also fuel is drained from the carburetor's fuel bowl and mixed with the air coming in to form a combustible combination.

At idle, the fuel goes into the carburetor throat through one or small idle ports just above the throttle plate. And at higher engine speeds, fuel is drawn through the main metering jets into the venturi (the venturi is the narrowest part of the carburetor throat). The air/fuel mixture will then flow down through the intake manifold and into the cylinders where it is burned to produce power.

Even though the main work of a carburetor is fairly simple, it also depends on a number of add-on devices for cold starting, idle control and emissions.

The changes in emission regulations around the early 1980s what was made carburetors not to be used anymore by motorists, as they were unable to meet up with the new emission requirements. When it got to the mid-1980s, carburetors became obsolete and were history on new production vehicles.

It was replaced by throttle body and multiport electronic fuel injection systems.

Being that the carburetor is the component of the vehicle that provides the fuel and air needed for the engine to run, the carburetor is one of the most important components on any vehicle that has one.

And if the carburetor develops a problem, the drivability and performance of the vehicle can be greatly affected. In this book, you will be equipped with the knowledge of common carburetor problems, how to diagnose what the problem with a carburetor is and common tips on how to fix these problems.

Chapter 1: Carburetor Problems

When a carburetor is bad or failing it will usually produce a few symptoms that can make the driver know that the attention may be required.

Some of these common signs include: reduced engine performance, black smoke coming from the exhaust, backfiring, overheating or hard starting. Let us look at these signs in more details.

Reduced Engine Performance

One of the first symptoms commonly found with a bad or failing carburetor is a reduced engine performance. The carburetor being the main tool in the vehicle that is responsible for metering and delivering the air fuel mixture needed for the engine to run, must be given special attention to.

And if it develops any sort of issue, it can throw the mixture off and affect the engine performance.

When a carburetor is faulty, it may cause the engine to have slow acceleration, reduction in power and fuel efficiency.

Black Smoke from Exhaust

When black smoke starts coming from the exhaust it is an indication that the engine is running excessively rich, (that means it is using too much fuel). A carburetor that is delivering too much fuel to the extent of producing black smoke is actually burning fuel unnecessarily, and producing excessive emissions that are not needed.

Back firing or overheating

Backfiring and overheating of engine is another common symptom of a potential problem with the carburetor. If a fault arises with the carburetor that makes it to start delivering a lean mixture, (a lean mixture is one that does not have enough fuel) it may cause the engine to start backfiring or overheating.

Apart from causing backfire and overheating, lean mixtures will also affect the engine performance and in critical cases, it can completely damage the engine

Hard Starting

When a carburetor is bad or failing, it can have a problem of hard starting. The air fuel ratio that the carburetor is responsible for blending and metering is very important and sensitive during cold starts.

So if the carburetor develops any form of issue that disturbs the air fuel ratio, it may end up causing "hard starting".

Based on the particular type of the issue, the vehicle may become very hard to start and then it gets better as the engine warms up, or it may still be difficult to start even when warmed up.

In case your vehicle is showing any of the symptoms mentioned above or you suspect that your carburetor may be developing a problem, take the vehicle to a professional technician to diagnose and determine if your vehicle carburetor needs to be serviced or replaced.

Chapter 2: Carburetor Problems

If your carburetor is clean and is working well, the engine should start with ease (whether the engine is hot or cold). It should idle smoothly, and also accelerate without stumbling. Also when the carburetor is clean, the engine should get normal fuel economy and emissions should be within limits.

The problems that arise when a carburetor is "bad" or "dirty" include: hard starting, hesitation, stalling, rough idle, flooding, idling too fast and poor fuel economy. To rebuild a carburetor can be quite tricky, and it is costly to replace, so be sure you know what exactly the issue with the carburetor is before you touch it.

Hard Cold Starting Problems

Hard starting can be as a result of a choke that fails to close and causes a rich fuel mixture when the engine is cold. In this case all you may have to do is simply to adjust or clean the choke mechanism and linkage.

You do not need to rebuild or replace the carburetor. Chokes are very sensitive components of a vehicle, and they can easily be misadjusted, that is why it has been recommended that auto makers, make choke and idle mixture adjustments "tamper-resistant".

Inside where the choke is housed is a round metallic heat-sensing spring that decreases in size when it is cool and enlarges (unwinds) when it gets hot. The spring is what is responsible for opening and closing the choke plate on top of the carburetor.

The spring can be found inside a black plastic choke housing on the top or the side of the carburetor. Also note that the spring is made hot by an electric heating element that is inside the cover and/or heat from the exhaust manifold that is drawn up into the housing through a small metal tube. In case the heating coil gets burnt or is not receiving voltage, or if the heat riser is plugged in with rust, if it is loose or missing, just know that the choke will not get warmed up properly.

This will make the choke to continuously stay on all the time, or too long, making the engine run rich and idle too fast.

With that said, if it happens that the bi-metal choke spring gets broken, you will find it difficult to close the choke. And when an engine is cold it normally requires a very rich mixture of air and fuel to start. So if the choke has developed a fault, it will take in too much air.

A broken choke will also not allow the engine to idle properly (no fast idle while it is warming up) and this can cause it to stall until it gets to the normal operating temperature.

The shaft is what opens and closes the choke. If it gets dirty, it may cause the choke to stick. It is the same with the choke linkage if it is dirty or damaged.

All that is needed when a choke is damaged is a choke repair kit or a new bi-metal spring to repair the starting problem.

No need to replace the whole carburetor. It is like replacing the engine just because the water pump is bad.

Hard starting could also be as a result of: vacuum leaks, problem with the ignition (worn out or dirty spark plugs, bad plug wires, cap, rotor, etc.) low compression, if the starter or battery has become weak.

Hard Hot Starting Problems

For hot starting problems, the carburetor may not really be the cause. A hot start problem is mostly caused by too much heat in the area around the carburetor, fuel lines or fuel pump.

The heat will cause the fuel in the fuel lines, carburetor bowl or pump to boil. And this will cause a "vapor lock" condition to be created which can make a hot engine difficult to start. In this case, there is no need to replace or rebuild the carburetor because the real issue is heat.

So all you need to do is to readjust the fuel line away from the source of the heat (like the exhaust manifold and pipe), and/or to insulate the fuel line by creating a heat shield or wrapping the fuel line with insulation.

Hot start problems can also arise as a result of too much resistance in a starter, poor battery cable connections, or a bad ignition module that misbehaves when it is overheated.

Hesitation or Stumble When Accelerating

Hesitation is a special symptom of a lean fuel mixture (i.e. too much air, not enough fuel) and it can be caused by either a dirty or a misadjusted carburetor, or one with a weak accelerator pump or worn out throttle shafts. In this case you may need to rebuild or replace the carburetor.

When the throttle is open, the accelerator pump will release an extra amount of fuel into the throat of the carburetor.

This helps to put away the extra air that has sneaked in until fuel begins to flow through the metering circuits and can meet up with the change in air speed through the tool called venturi (the venture is the narrow part of the carburetor throat).

The accelerator pump can either use a rubber diaphragm or a rubber cup on a piston to pump fuel through its discharge nozzles. If the diaphragm gets torn or the piston seal becomes worn out, the accelerator pump may not convey its normal amount of fuel. Also if the discharge nozzles are plugged in along with dirt or fuel varnish deposits, it can limit the flow of fuel.

To check how the accelerator pump operates, you can remove the air filter, looking down into the carburetor, and pumping the throttle. Observe carefully to see if a jet of fuel will squirt into each of the front venturis (barrels) of the carburetor.

If you don't see fuel squirting out, or the stream is very weak, or only one of the two discharge nozzles on a two-barrel or four-barrel carburetor are working, just know that the accelerator pump circuit has a problem.

Usually fuel enters the accelerator pump through a one-way steel check ball. The ball allows fuel to get in, but it will get pushed back against its seat by pressure inside the pump when the throttle opens. So if this check ball is stuck open, it will act like a pressure leak and would stop the accelerator pump from squirting fuel through the discharge nozzles.

Also if the check ball is stuck shut, it will not allow fuel to enter the pump and there will be no fuel to pump through the discharge nozzles. If fuel varnish deposits, enters and occupies the carburetor jets, or dirt enters inside the fuel bowl, the flow of fuel can be restricted and it will cause a lean condition.

The possible solution to this is to clean the carburetor with carburetor cleaner. This will get rid of the dirt and fuel varnishing deposits. Cleaning it will also enable the carburetor to be restored to its normal operation.

If air leaks somewhere else on the engine it can also lean out the fuel mixture. The possible means by which air can come in to the intake manifold is through loose or cracked vacuum hoses, emission hose or the PCV system.

The vacuum leak in the carburetor base gasket or insulator, or the leak in the intake manifold gaskets, power brake booster or other vacuum accessories can release unwanted air. Another source through which air can enter into the manifold is badly worn valve guides and seals

Another cause of hesitation could be if the EGR valve fails to close at idle or when the engine is cold.

Other causes of hesitation may include a damaged distributor advance mechanism, a weak ignition coil, if there are carbon tracks on the coil tower or the distributor cap, damaged plug wires, worn out or dirty spark plugs that misfire when the engine is under load, or even an exhaust restriction. Also bad gas can cause hesitation problems.

So before you think of rebuilding or replacing the carburetor, these other possibilities need to be checked and taken out.

Hesitation Under Load

A hesitation, stumble or misfire occurs when the engine is under load can be caused by a bad power valve inside the carburetor. Carburetor uses intake vacuum to siphon fuel through its metering circuits.

When the engine load increases and the throttle enlarges, the intake vacuum will drop. As a result of this, the flow of fuel can reduce and this will make the fuel mixture go lean. So take note of the power valve which has a spring-loaded vacuum-sensing diaphragm.

This opens to increase the flow of fuel when the vacuum drops. If the diaphragm has failed or the valve is filled with dirt particles or fuel varnish deposits, it must be replaced. A carburetor rebuild kit usually comes along with a new power valve.

Weak ignition oil can also cause hesitation or misfiring under load. Also cracks in the coil or distributor cap, bad spark plug wires can be a cause of hesitation or misfiring under load.

Stalling

What makes an engine to stall is if the idle speed is too low. Other causes are if the fuel mixture is too lean, if it won't burn, or if it stops flowing or the ignition system runs out of spark.

To fix the stalling problem is not by rebuilding or replacing the carburetor as that won't solve it.

Rather if stalling is ignition related or due to a weak fuel pump, plugged fuel filter or fuel line, or bad gas (too much water or alcohol) a simple adjustment is all that would be necessary to increase the idle speed or make the idle mixture rich.

But if the engine is taking in air through a vacuum leak somewhere, adjustment may completely wipe out the possibility to stall.

You must find the vacuum leak and fix it before proper idle speed and mixture adjustments can work.

You may have to rebuild and replace the carburetor if there are internal air leaks in the carburetor itself, a sticky needle valve would be starving the carburetor of fuel, or the jets, the air bleeds or metering passageways in the carburetor may be dirty or plugged. There may be a need for replacement if the throttle shafts are badly worn, or the carburetor housing is warped or damaged.

When it has to do with vehicles with computer-controlled idle speed, an inoperative or defective idle speed control (ISC) motor can make an engine stall.

So, what is the function of the ISC motor?

The ISC motor function is to maintain the needed idle speed by properly aligning throttle linkage. So a bad electrical connection or wiring problem can prevent the motor from carrying out its function.

If you observed that the ISC motor is receiving voltage and is properly grounded but does not move, then know that the motor is burned out and needs replacement. The motor may have failed to work because a vacuum leak caused it to overwork itself in a fruitless attempt to compensate for the unwanted air.

Rough Idle

Rough idle problems are commonly caused by an overly lean fuel mixture that results in lean misfire.

Another common cause of idle problems is air leaking between the carburetor and intake manifold. The solution is to tighten the carburetor base bolts or change the gasket under the carburetor. Air usually leaks in vacuum lines or in the PCV system or EGR valve.

Other causes that are carburetor-related includes:

an idle mixture adjustment set too lean (what you should do is to back out the idle mixture adjustment, then screw one quarter of a turn at a time until the quality of the idle improves), or a dirty idle mixture circuit (which may require cleaning and rebuilding the carburetor).

Other possible causes of a rough idle includes: a bad charcoal canister purge control valve that does not close and is leaking fuel vapors back into the carburetor, excessive compression blow by (worn rings or cylinders), weak or broken valve springs, or ignition misfiring due to worn or dirty spark plugs, bad plug wires or a weak ignition coil.

Flooding

Flooding is a vehicle problem that is mostly but not always caused by the carburetor.

The carburetor may get flooded if dirt particles enters the needle valve and stops it from closing. Since no way to shut off the flow of fuel, the bowl will overflow and pour fuel into the carburetor throat or out of the bowl vents. And once an engine is flooded it may not start because the plugs are wet with fuel.

CAUTION: take note that flooding can be a very dangerous situation because it will create a serious fire hazard if fuel pours out of the carburetor onto a hot engine.

Flooding can also occur in the carburetor if the float inside the fuel bowl is set too high or if it develops a leak and sink (this is applicable to hollow brass or plastic floats majorly). All you need may just be a new float, thus there would really be no need to replace the whole carburetor. Floats do not come along with a rebuild kit, so if new gaskets are required, you will have to purchase a rebuild kit.

Excessive fuel pressure forcing fuel to pass through the needle valve can also be a cause of flooding. Excessive heat is another factor that causes flooding. A heat riser valve if it's on a V6 or V8 engine that sticks shut can make a hot spot under the intake manifold. And that will make the fuel in the carburetor to start bowling to start boiling over to the extent of flooding the engine.

Idles Too Fast

The idle problem commonly caused by the automatic choke. If the choke is sticking, the engine will remain constant at fast idle for too long. Do a proper check on the choke and the choke linkage, clean or repair it, whichever is needed.

When you inspect your vehicle, you will observe that there is a separate fast idle adjustment screw on the choke linkage that is responsible for controlling the engine speed as the engine is warming up.

You will observe that the tip of the screw rests against a cam that slowly rotates as the choke opens during engine warm up. You are to turn this screw anti-clockwise to reduce the fast idle speed, or clockwise to increase the fast idle speed.

A high idle speed can also be as a result of vacuum leaks. This will allow air to enter the manifold (leaky PCV hose, power steering booster hose or other large vacuum hose). Another cause of a high idle may be as a result of a defective ISC motor stuck in the extended (high idle speed) position.

Poor Fuel Economy

The carburetor may not be the fault in this case. The real problem may just be a lead foot on the accelerator pedal, or maybe the engine has low compression, retarded ignition timing or an exhaust restriction (plugged converter). Apart from all these, the carburetor may have a misadjusted or heavy float, or the metering jets may be too large.

The float setting is what determines the level of fuel in the bowl, this in turn affects the richness of the Air/Fuel mixture.

A float that is set too high or has become filled in with fuel (this problem continues to affects many foam plastic floats today), will allow the fuel level to rise and make the fuel mixture rich.

To determine this condition, you need to check the float level and the float needs to be weighed to ascertain if it has become saturated with fuel. If you find out that the float is heavy, then it has to be changed.

Using electronic feedback carburetors, a slow or completely damaged oxygen sensor can richen fuel mixture. So also a defective coolant sensor will never allow the feedback system to go into closed loop. Checking for fault codes and sorting out the operation of the feedback system can eliminate these possibilities.

If the carburetor is newly changed with a used carburetor or a carburetor of another engine, the jets may not be calibrated rightly for the new application. Jets that are bigger usually flow in more fuel and richen the fuel mixture.

By installing jets of a smaller size, you may be able to restore the proper air/fuel mixture and be able to manage fuel.

A proper way to determine if the fuel mixture is too rich or too lean is to investigate the spark plugs. If you observe that the plugs have heavy black, sooty carbon particles on the electrodes, know that the fuel mixture is too rich.

To determine if the mixture is too lean, you will observe that the ceramic insulator around the center electrode may look yellowish or blistered in appearance. An air/fuel mixture that is too lean is bad because it can damage the engine pre-ignition and detonation.

Chapter 3: Possible Solutions

If the carburetor needs to be worked on, you can either rebuild it with a kit or replace it with a new or remanufactured carburetor. To replace a carburetor is expensive, and the price varies based on the application and type of carburetor.

Another option is also to clean and rebuild an older one or two barrel carburetor which is a quite simple to do.

But take note that four barrel may be more difficult clean. Complicated carburetors e.g those that have a variable-venturi or electronic feedback controls and tamper-resistant adjustments can also be difficult to rebuild, and they definitely require the skills of an expert which may be costly too.

So it is advisable to replace a more complicated carburetor than to attempt to rebuild one, as it is easier and less risky to replace.

In the instance that the carburetor has worn out throttle shafts, which is causing air to leak out, or if any of the castings are cracked, warped or damaged, just know that the carburetor cannot be rebuilt, instead it must be replaced. The only option is if you have another carburetor you can scout for parts to help and repair the first carburetor.

Whichever it is that you want to do, whether rebuilding or replacing a carburetor, you first need to identify it. The year, the type, model and engine size is not enough information to find the correct carburetor kit or replacement carburetor.

A small metal ID tag usually comes along and can be found on the carburetor. The tag will give you the exact model number and calibration of the unit.

Chapter 4: Carburetor Rebuilding Tips

Before you start dismantling the carburetor, first search for an assembly diagram in a service manual for easy reference and assistance. The Carburetor kits may either come or may not come along with an assembly diagram and instructions.

Take note of where various vacuum hoses and lines make a connection to the carburetor. Possibly, draw a picture of the hose connections, or you can place a piece of masking tape on each hose and write on the tape to remember which hose goes where.

Display the parts of the carburetor on a clean work bench, on a paper or a metal tray. Observe closely how the parts came apart (especially linkages) this will help you to remember how to reassemble the parts when you put the carburetor back together. Check for small steel check balls that can be easily get missing.

While cleaning a carburetor parts, always use a carburetor cleaner or a solvent that will not destroy the plastic and soft metal parts. Always wear rubber gloves on both palms to prevent the skin from having contact with the cleaner or solvent. Adhere to the instructions for use on the cleaner or solvent, and ensure to use it in a well-ventilated area. Do not inhale the fumes.

Look out for a worn throttle shaft. The hole in the base casting after some time can become worn out, and it will start allowing air to get in through the shaft. And this will make the fuel mixture to lean out, and most likely cause lean misfire, hesitation or stumbling problems. If the throttle shaft hole is worn, you can fix it by removing the throttle shaft, or by drilling out the hole to oversize and also by installing a steel or brass sleeve to restore it to clear normally.

Other issues to check for, is a bad float inside the fuel bowl. If it is a brass float, shake it vigorously to find out if there is any liquid substance inside.

Also having a small hairline crack in the seam can make the fuel to seep into the float, this will cause it to sink and it will flood the engine with too much fuel.

A lot of carburetors have plastic floats instead of brass. And some of these plastics, like sponge will soak up fuel with time, making them to become too heavy. The float will now begin to ride too low in the fuel bowl and will flood the engine with excessive fuel. The solution to this is to replace the float with a new one (that is if you can find a replacement).

Chapter 5: Carburetor Installation Tips

The carburetor mounting surface on the intake manifold should be cleaned (DON'T allow any form of dirt or gasket debris to fall down inside the manifold), and then install a new base gasket under the carburetor (don't ever reuse the old gasket because at most times they are already leaking).

You can use gasket sealer and apply it to the base gasket to limit the chance of its getting air leakage. Never use RTV silicone because it immediately dissolves when it gets exposed to gasoline.

Make sure the carburetor base mounting nuts or bolts gets tightened evenly so that the gasket gets firmly clamped in place. Don't tighten the fasteners too much. Doing so can warp or crack the carburetor base plate.

When you are reconnecting the fuel line and any other fittings (EGR, PCV) to the carburetor, take proper care not to cross-thread the fittings, and never over-tighten it, doing so can strip the treads in the soft casting.

A new fuel filter should be installed to prevent dirt from entering the carburetor.

NEVER forget to reattach the throttle return spring(s) on the throttle linkage when you finish. Else what you get is a runaway engine when you start it up. Replace the springs with new ones if they are old and have rusted, and if they appear to be stretched or are weak.

Make sure the throttle opens all the way when the gas pedal is floored by testing the throttle linkage, and ensure that nothing binds or rubs against the linkage as that may cause it to stick.

When you are installing the air cleaner, make sure you DON'T over-tighten the nut which holds the air cleaner together as this can displace and damage the carburetor casting.

Check all the rubber fuel hoses and clamps. Any hose that is hard must be replaced.

So also any one that is brittle, mushy, cracked or leaking. It is also recommended for you to acquire new clamps. Worm-screw clamps are recommended as the best. There are others like the ring style clamps, but this one loses tension with age, it can also get deformed permanently if they are over-expanded when they are being removed.

Check and recheck all the fuel line, check the vacuum and emission hose connections, also check the throttle linkage and return spring, then start the engine. Check again for any leaks or other problems.

Chapter 6: Carburetor Adjustment Tips

Once the engine gets to its normal operating temperature, adjust the idle speed and idle mixture adjustment screws. The idle speed should be set to specifications (which is 550 to 650 rpm), also adjust the idle mixture screws for smoothest idle. Each idle mixture screw must be turned in until the engine starts to stumble, then back it out to about 1/4 to 1/2 turn.

If the engine does not start with ease, you may have to adjust the automatic choke. When the engine is cold, the choke should be fully closed, and it should be opened all the way once the engine has warmed up. Little adjustments would do a lot, and you may have to try and try several adjustments of the choke housing before you get it right.

If hesitation or stumbling occurs in the engine when accelerating, the accelerator pump linkage or cam may need some adjustment in order to increase the quantity of fuel squirted into the engine when the throttle opens. The accelerator pump linkage or cam always has several adjustment settings, so attempt the next higher setting if it needs more fuel.

In case you are installing a performance carburetor, the main metering jets that come along with the carburetor probably give you or may not give you the best air/fuel mixture. You can achieve the best performance with a slightly rich mixture.

Indications are usually on jet sizes. Such indications come with a number stamped on the side of the jet. When you install a jet that is not too large, more fuel and richen mixture will flow. In the circumstance that the carburetor is running with too rich mixture, switch to a not-too-small sized jets, this may give a better performance.

To replace the main metering jets means removing the top of the carburetor or the fuel bowls. However some racing carburetors have jets that can be replaced without dismantling the parts.

Chapter 7: Categories of Carburetor problems

Carburetor problems have been categorized and highlighted with their possible solutions below for easy understanding:

Problems

•   When the Engine hunts (at idle or high speed)

•   If the carburetor comes out of adjustment

•   If the engine fails to start

•   If the engine is failing to accelerate

Possible Solution

If you notice any of the above problems, examine the idle and main mixture adjustment screws and also examine the O-Rings if there are any cracks

Problems

•   If the engine hunts (at idle or high speed) or

• Your engine refuses to idle

• If the engine lacks power when it is at high speed

• If the engine is over speeding

• And if the engine lacks for fuel at high speed (leans out)

Possible Solution

The possible solution would be to adjust the main mixture adjustment screw; some models of this screw demands finger tight adjustment.

Problems

• If the carburetor goes out of adjustment

• Instance where the engine refuse to start

• Or if the engine fails to accelerate

• Problem of engine hunts (at idle or high speed)

• Also if the engine refuses to idle

- If you notice that your engine lacks power when it is at a high speed

- And if you observe that the Idle sped is excessive

Possible Solution

What you should do, is adjust the idle mixture screw

Problems

- In case the carburetor goes out of adjustment

- If the engine fails to idle

- And if idle speed is excessive

Possible Solution

Possible solution is to look out for any bent choke and throttle plates

Problems

- Failing to start Engine

- If the engine hunts (at idle or high speed)

- If the engine also fails to idle

- Cases where the engine over speeds

- If the Idle speed is too much

- And when the engine lacking fuel at high speed (if fuel leans out)

Possible Solution

Try to adjust the control cable or linage, this will ensure that there is full choke and carburetor control

Problems

Other possible problems are:

- When the Carburetor goes out of adjustment

- If the engine fails to start

- When the engine is over speeding

Possible Solution

Try to clean the carburetor after you have removed all the non-metallic parts that can be serviced

Problems

- Problem of carburetor getting flooded

- When the engine lacks fuel at high speed (leans out)

Possible solution

Inspect the inlet needle and the seat to know the condition be sure the installation was done properly.

Problems

- If the carburetor starts to leak

- Or if the Engine is over speeding

- Or if the Idle speed is too much

Possible solution

Inspect the sealing of the Welch plugs, caps, plugs and gaskets

Problems

• Instances where the carburetor comes out of adjustment

• If the engine is refusing to idle

• Where the engine power fails at high speed

• Or if the carburetor is leaking

• and where you observe that the engine is over speeding

Possible solution

Simply adjust the governor linkage

Problems

• If the engine hunts (at idle or high speed)

• Where the engine will not idle

- Or if the engine is lacking power when it's on a high speed

- Also if the carburetor gets flooded

- Or where the engine is lacking for fuel and it's on a high speed (leans out)

Possible Solution

Just adjust the float settings if it's a float type carburetor.

Problems

- Where the engine hunts (at idle or high speed)

- Or if the engine is refusing to idle

- If the carburetor gets flooded

Possible solution

Also adjust the float setting if it is a float type carburetor.

Problems

- Engine will not start

- Engine will not idle

- Engine lacks power at high speed

- Carburetor floods

- If idle speed is excessive

Possible solution

Inspect the diaphragm to see if there are cracks or any distortion and check the nylon check ball for function if it is available.

Problems

- If the engine hunts (at idle or high speed)

- Or the engine is lacking power when it's at a high speed

- Or if the carburetor gets flooded

- Where you notice that the idle speed is too much

- Or the engine is lacking for fuel at high speed (leans out)

Possible Solution

The possible solution is to check the sequence of gaskets and diaphragm to know the particular carburetor being repaired

CONCLUSION

Make sure you carry out consistent inspection of the carburetor parts. Always check your owner's manual for specific carburetor tuning and instructions. Inspect the float shaft for wear and check the float for leaks or dirt particles.

Before you service or repair any power equipment, first disconnect the spark plug and battery cables to avoid electrocution. Always remember to wear the appropriate safety glasses for your eyes and gloves for your hands to protect you from harmful chemicals and dirt particles.

www.ingramcontent.com/pod-product-compliance
Lightning Source LLC
Chambersburg PA
CBHW031501210526
45463CB00003B/1024